Now You Know Science

Move It!

Honor Head

W

FRANKLIN WATTS

LONDON • SYDNEY

First published in 2009 by Franklin Watts

Franklin Watts
338 Euston Road, London NW1 3BH

Franklin Watts Australia
Level 17/207 Kent St, Sydney, NSW 2000

Copyright © Franklin Watts 2009

Created by Taglines Creative Ltd: Jean Coppendale and Honor Head
Design: Paul Manning
Consultant: Terry Jennings

ISBN: 978 0 7496 8724 3

Dewey classification: 531

A CIP catalogue for this book is available from the British Library.

Picture credits
t=top b=bottom l=left r=right

3, 21, Tyler Olson, Shutterstock; 5, 26, 30br Hallgerd, Shutterstock; 6, 29t Ieva Geneviciene,
Shutterstock; 7, 29b Sonja Foos, Shutterstock; 8 Timothy Geiss, Shutterstock; 9, 29 Goh Siok Hian,
Shutterstock; 10 Mandy Godbehear, Shutterstock; 11 Kaspars Grinvalds, Shutterstock;
12 Greenland, Shutterstock; 13 Sandy Maya Matzen, Shutterstock; 14 Juriah Mosin, Shutterstock;
15, 29b Vuk Vukmirovic, Shutterstock; 16 Losevsky Pavel, Shutterstock; 17 Anton Gvozdikov,
Shutterstock; 19 Polushkin Ivan Nikolaevich, Shutterstock; 20 Alpenglow Design LLC, Shutterstock;
22 Drazen Vukelic, Shutterstock; 23 Maxim Petrichuk, Shutterstock; 24b Matka Wariatka,
Shutterstock; 24t Bezmaski, Shutterstock; 25b Mindy w.m. Chung, Shutterstock;
25t Raia, Shutterstock; 27, 30l Darren Baker, Shutterstock.

Printed in China

Franklin Watts is a division of Hachette Children's Books, an Hachette Livre UK company.
www.hachettelivre.co.uk

Contents

What are forces?

What happens when you push a buggy?

It moves forward.

◀ This man pushes the buggy to move it forwards.

Pushes and pulls are called forces.
Forces make things move.

▲ This girl is using the handle to pull
the trailer behind her.

Pushing

What do you have to do to make a swing move?

▲ Nothing will move without forces.

Push it! When you push a swing it moves forwards.

▲ This little girl gives her brother a push forwards. What happens next?

Pulling

These boys are playing tug of war. They are pulling on either end of the rope.

▲ What would happen if one boy suddenly let go?

This steam engine is pulling the train carriages along behind it.

carriages

steam engine

▲ Without the engine, the carriages would not move.

Making it move

There are lots of ways you can make things move. When you ride a bike, you push down on the pedals to make the wheels turn.

▲ You have to push the pedals to make the bike move forwards.

You pull on the handlebars
to change direction.

Handlebars turn
the front wheel.

To turn a corner,
you have to point
the wheels in a
different direction.

13

Light and heavy

A light weight needs only a small push to move it.

▲ A small kick will make this light ball move a long way.

A heavy weight needs a big push to move it.

This heavy car will only move very slowly.

STOP!
You may hurt yourself if you try to push a heavy object.

Rolling along

Wheels help us to move very heavy objects.

It would be hard to push or pull this suitcase if it didn't have wheels.

On a flat surface, it is easier to lift or roll a heavy object than it is to slide it.

▲ This heavy box needs a fork-lift truck with wheels to move it.

Nice and smooth

It is easier to move an object on a smooth surface.

▶ Skates move easily over the smooth ice.

A bowling alley has a smooth surface for the ball to roll down.

▲ Could this ball roll down a rough, bumpy surface?

Going faster

It is easier to move something down a hill than to push it uphill.

▲ Sliding downhill is fun – walking up is hard!

The steeper the slope, the faster you go!

▼ Long, smooth skis help you to whizz down a mountain slope very quickly.

skis

Slow down... stop!

Once something is moving, it will keep going until a force slows it down or stops it.

◀ Air pushes against the parachute and slows the man down.

The brakes are used to stop a bicycle. They rub against the wheels to stop the bike moving.

You have to pull on the brakes very hard to stop a bike when it is going downhill.

Changing shape

Forces also help things to change shape. Blowing and squeezing are forces.

▶ What happens when you blow into this flat balloon?

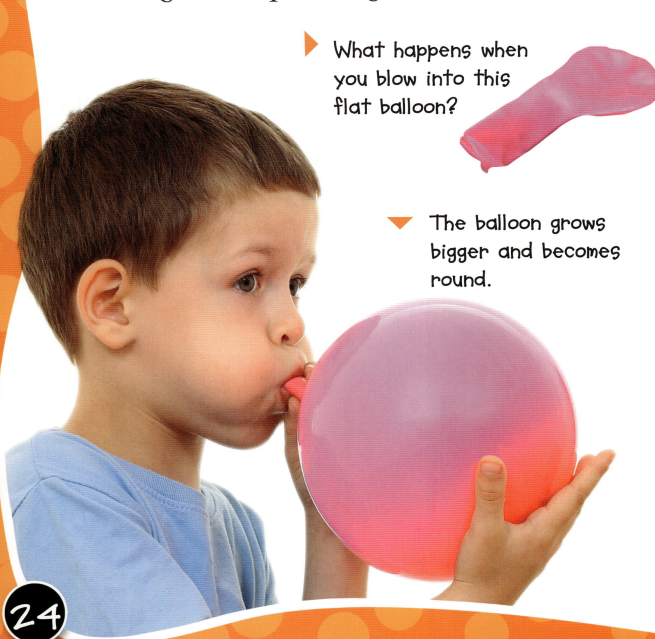

▼ The balloon grows bigger and becomes round.

▶ What happens when you squeeze this sponge?

▼ The sponge changes shape when you squeeze it.

Using wind

Wind is moving air. When you fly a kite, you need the wind to lift the kite into the air.

The children are running to catch the wind in their kite.

Wind can also push things along.

▶ The wind blows the boat's sails and pushes the boat along.

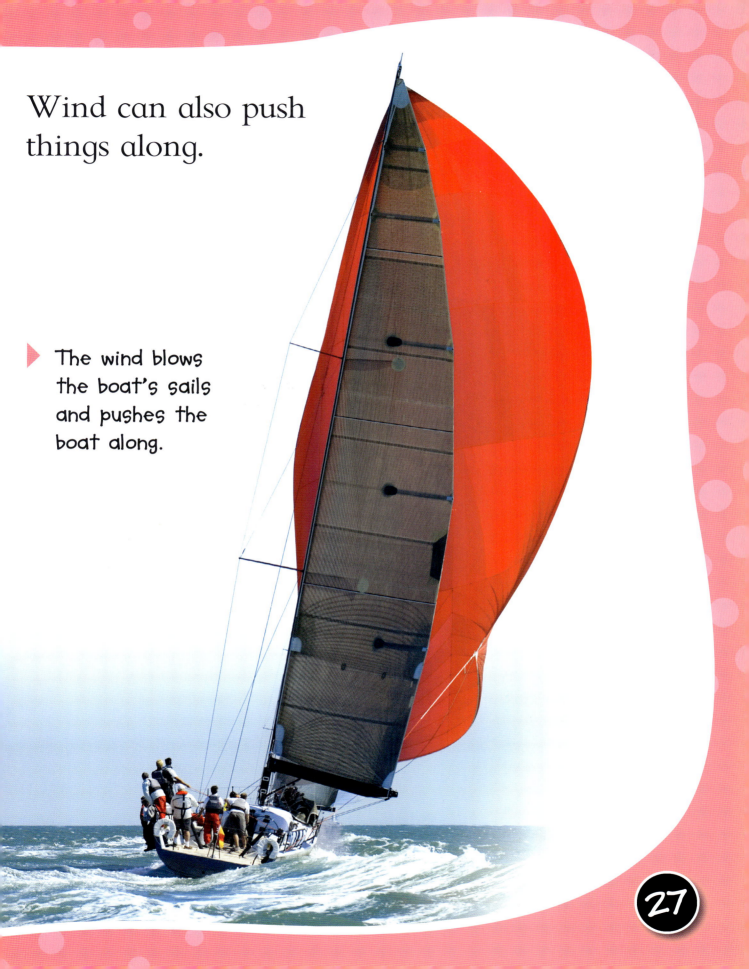

Things to do

Odd one out

Which picture does not show pushing?
What force does it show?

a

b

c

d

Watch out – wind about!

Which pictures show how the wind makes things move?

a

b

c

Talk back

Have you used any forces today?
Have you pushed anything?
Have you pulled anything?
What other forces have you used?

Glossary

carriages Part of a train where the passengers sit.

direction The way in which you want to go.

fork-lift truck A small truck used to pick up heavy loads and move them.

smooth Very flat and even.

Index